Australia's Animals & Wildflowers

Australia's Animals & Wildflowers

WELDON RUSSELL
PUBLISHING

Other titles in the series:
Australia Images of a Continent
Australia's Outback

First published in Australia in 1992 by Weldon Russell Pty Ltd
107 Union Street North Sydney NSW 2060 Australia

A member of the Weldon International Group of Companies

Copyright © 1992 Weldon Russell Pty Ltd

Publisher: Elaine Russell
Managing editor: Dawn Titmus
Senior editor: Ariana Klepac
Project coordinator: Margaret Whiskin
Picture researcher: Anne Nicol
Captions: Anne Matthews
Design concept: Catherine Martin
Designer: Jean Meynert
Paste-up artist: Megan Appleby
Production: Jane Hazell, Di Leddy

All rights reserved. No part of this publication may be reproduced, stored in a retrieval system, or transmitted in any form or by any means, electronic, mechanical, photocopying, recording or otherwise, without the prior written permission of the copyright owner.

National Library of Australia Cataloguing-in-Publication data

Australia's animals and wildflowers.

ISBN 1 875202 50 1.

1. Botany - Australia - Pictorial works. 2. Zoology - Australia - Pictorial works.

574.994

Produced by Tien Wah Press, Singapore

A KEVIN WELDON PRODUCTION

RIGHT *A group of long-necked brolgas* (Grus rubicunda) *make their way across a Northern Territory waterway. These birds are the only crane native to Australia.*

FRONT COVER *A crimson rosella* (Platycercus elegans) *enjoys the seeds of a banksia flower. These brilliantly coloured members of the parrot family are abundant in forest and coastal scrub areas.*

BACK COVER *The kookaburra* (Dacelo gigas) *is our most familiar member of the kingfisher family and an enduring symbol of Australia—it is unique to the continent.*

ENDPAPERS *Detail of a sunflower* (Helianthus annuus). *These plants thrive in Australia and are grown commercially for their oil-rich seeds. They often reach 1.75 m in height and the flower heads can be up to 30 cm across.*

HALF TITLE PAGE *The long-necked pied or magpie goose* (Anseranas semipalmata) *frequents swamps and floodplains of the northern region of Australia.*

OPPOSITE TITLE PAGE *Found only in the Daintree and other rainforests of eastern Queensland, the tree-dwelling Boyd's forest dragon* (Gonocephalus boydii) *is regarded as a potentially endangered species.*

TITLE PAGE *This small, dew covered plant* (Trianthema megasperma) *is found in the Kakadu region.*

OPPOSITE CONTENTS *The snow daisy* (Celmisia longifolia) *is a perennial herb found in the Alpine regions of Australia.*

CONTENTS *There are some 45 species of kangaroo in Australia, including wallabies and the familiar larger kangaroo, like this attractive specimen.*

Contents

Mammals 9

Trees 21

Birds 33

Wildflowers 47

Reptiles, Insects and Frogs 59

Mammals

MAMMALS

PREVIOUS PAGES *The spiny anteater, or echidna (Tachyglossus aculeatus), is one of only two egg-laying mammals—along with the platypus it forms the special category of monotremes, regarded as the lowest order of mammal. Echidnas are covered with sharp quills which provide their only protection from other animals.*

LEFT *Australia's familiar and often maligned feral dog, the dingo (Canis familiaris dingo). Dingoes hunt for their food at night and spend most of the day in caves or other hiding places.*

BELOW *Camels were introduced to the continent in the mid-1800s and are well-suited to their Australian outback environment. It is believed that there may be as many as 200,000 feral camels in this country, all of them one-humped dromedaries like these Simpson Desert inhabitants.*

MAMMALS

LEFT *This fine example of a red kangaroo (Macropus rufus) was pictured in South Australia's Flinders Ranges. The red variety is widespread throughout the continent and is the largest of Australia's marsupials.*

BELOW LEFT *The long-nosed bandicoot (Perameles nasuta) is a nocturnal marsupial with sharp teeth and three claws on each front paw. These claws are used to extract earthworms, bulbs and plants from the ground.*

BELOW RIGHT *Pademelons are macropods, or herbivorous marsupials, that resemble kangaroos and wallabies. These smaller relatives prefer the vulnerable rainforest environments of the eastern seaboard and their numbers have subsequently suffered a reduction due to land clearing in these areas.*

MAMMALS

LEFT *A familiar koala's sleepy pose! The tree-dwelling koala (Phascolarctos cinereus) inhabits only the eastern and far southern regions of Australia and feeds predominantly on only a few varieties of eucalyptus leaves. These fussy eating habits, as well as disease, have contributed to a substantial and alarming decline in koala numbers in recent years.*

RIGHT *The stout, sturdy wombat is undoubtedly one of Australia's favourite marsupials. This common wombat (Vombatus ursinus) forages in the snow at Kiandra, New South Wales, for grasses and roots which form part of its herbivorous diet. Wombats can weigh up to 36 kg and have lived for as long as 27 years in captivity.*

Trees

TREES

PREVIOUS PAGES *A classic outback scene with a creek and its accompanying shading eucalypts, captured in northern Queensland. The eucalypt is Australia's most dominant tree with over 450 defined species which are spread across the continent from Tasmania to Cape York.*

LEFT *Detail of a snow gum in the Snowy Mountains. The bark of these trees is shed annually to reveal a white trunk, streaked with greys and greens.*

RIGHT *Snow gums (Eucalyptus pauciflora) are abundant in the high country of south-eastern Australia. These hardy trees flourish at higher altitudes, even above 1800 m, despite the associated cold temperatures and harsh conditions.*

TREES

Despite Australia's proliferation of eucalypt varieties, many introduced trees are used for specific purposes. This plantation of poplar (Populus), a common northern hemisphere native, at Echuca in Victoria is used for matchwood production.

The continent's eucalypts are many and varied. This remarkable, hardy tree has adapted to survive in conditions as varied as high alpine plateaux, semi-desert and, as this scene reveals, swamps and other water environments.

LEFT *The ghost gum (Eucalyptus papuana) is widespread in northern and central Australia. These trees have a smooth, pale bark and their whiteness is further enhanced by the presence of a fine dust which coats the trunk.*

RIGHT *A eucalypt at Ormiston Gorge in the Northern Territory. This hardy eucalypt prefers sandy soils and thrives in arid, rocky locations that would not suit other varieties.*

BELOW *A close look at the trunk of this eucalypt in the Victorian Alps reveals a fascinating pattern of shapes and colours in its bark.*

TREES

LEFT *In the arid north-west, the twisted white branches of a solitary gum tree form a striking contrast to its rocky, spinifex-covered surroundings.*

RIGHT *The indigenous bottle-shaped baobab* (Adansonia gregorii) *is found mainly in north-western Australia. The excessively thick trunks of these deciduous trees can reach 24 m in circumference.*

TREES

TREES

OPPOSITE PAGE *A magnificent river red gum* (Eucalyptus camaldulensis) *dominates an upland region of South Australia's Flinders Ranges. Although these trees are frequently found near watercourses, they are capable of thriving at higher, more arid, altitudes.*

LEFT *Blackboys (genus* Xanthorrhoea) *in the Barossa Valley of South Australia. This unusual, slow-growing tree features long narrow leaves which resemble grass: hence the alternative and popular name of grass tree.*

Birds

BIRDS

PREVIOUS PAGES *A flock of airborne pelicans forms a striking black and white pattern as it crosses the Northern Territory's gulf country.*

LEFT *With an average length of 40–45 cm, the king parrot (Alisterus scapularis) is one of the largest members of the Psittacidae family which includes parrots, lorikeets and cockatoos. These magnificently coloured birds are best suited to an environment of large trees and are fairly common in coastal regions.*

BELOW *Crested pigeons (Ocyphaps lophotes) feature a distinctive erect black crest and are sometimes mistaken for the less common topknot pigeon. The crested variety is mostly found inland, but it prefers to be near water.*

ABOVE *Australia's largest bird of prey is the wedge-tailed eagle (Aquila audax) or 'bold eagle'. This magnificent creature is the world's fourth-largest eagle and has been known to have a wingspan of over 2.75 m. It is found throughout Australia but is more common in central, rather than coastal, regions.*

LEFT *The kookaburra (Dacelo gigas) is our most familiar member of the kingfisher family and an enduring symbol of Australia—it is unique to the continent. Also known as the laughing jackass because of its boisterous 'laugh' this handsome bird has a distinctive body and beak shape and is very common throughout Australia.*

BIRDS

Sadly, the attractive Major Mitchell cockatoo (Cacatua leadbeateri) is becoming increasingly rare. These beautiful delicately pink-toned, crested birds are now found only in isolated pockets in dry regions of the inland.

LEFT *Reaching a height of around 1.2 m, the jabiru* (Xenorhynchus asiaticus) *is Australia's only representative of the stork family. This long-legged, stately bird is also known as the black-necked stork and inhabits coastal areas, lakes, and swamps of the north and north-east.*

RIGHT *The graceful brolga* (Grus rubicunda), *Australia's only native crane, performs a characteristic mating dance which is believed to have had a substantial influence on Aboriginal dance movements. These large birds nest and breed in pairs and are common in the swampy areas of the continent's north.*

ABOVE LEFT *The strikingly marked regent bowerbird (Sericulus chrysocephalus) is somewhat less common than its satin cousin, but also inhabits rainforest regions and has the same tendency to create a bower in addition to its nest.*

ABOVE RIGHT *Although the peafowl (Pavo crisatus), a member of the pheasant and quail family, is an introduced species, it can survive and breed in the wild in Australia, such as on Rottnest Island in Western Australia. This peacock displays the typical, almost luminous, rich blue colouring of the species.*

RIGHT *All but two of the nine species of bowerbirds exhibit the peculiar habit of building a bower of sticks, leaves and grasses—a ritual which seems to be connected with mating. Satin bowerbirds (Ptilonorhynchus violaceus), pictured here in Queensland's Lamington National Park, collect blue flowers, feathers and other items to decorate the bower's entrance in the hope of enticing a mate.*

ABOVE *Australia's emu* (Dromaius novaehollandiae) *is, after the ostrich, the world's second largest bird and has been given a special place of honour on the nation's Coat of Arms. The flightless emu is common through most parts of the continent. It stands at around 1.5 m high and usually travels in flocks—it is a swift runner, capable of reaching speeds of up to 65 km/h.*

LEFT *The pelican* (Pelecanus conspicillatus) *is widespread throughout the continent, inhabiting both salt- and fresh-water environments. Its distinctive colouring and long, pouched bill make the fish-eating pelican one of our most recognisable and favourite birds.*

Wildflowers

WILDFLOWERS

PREVIOUS PAGES *The genus Acacia is represented by some 600 species in Australia, which are widely spread through the continent. This yellow wattle illustrates the typical 'fluffy' appearance of the flowers created by the dense arrangement of the numerous stamens.*

LEFT *Blue boys (Dampiera purpurea) grow in sandy soils in Queensland, New South Wales and Victoria. They feature attractive purple and yellow flowers that bloom during the spring and the summer months.*

BELOW LEFT *The sturt desert rose is the floral emblem of the Northern Territory. Gossypium sturtianum is a shrub that grows on stony ground and in arid conditions, yet somehow in this harsh environment it manages to produce these delicate pink, hibiscus-like flowers.*

RIGHT *The waratah (Telopea speciossima) is the state flower of New South Wales and a member of the Proteaceae family which includes grevilleas. This tall shrub grows only in New South Wales and produces its large, dramatic red flowers between September and November.*

WILDFLOWERS

Found in the lagoon regions of northern Australia, the lotus lily (Nelumbo nucifera) produces winter flowers which range in colour from red to pink, white, or even yellow. The plant's characteristically flat leaves can reach an amazing 100 cm in diameter.

LEFT *Kangaroo paws come in a variety of colours but the red and green one* (Anigozanthos manglesii) *is perhaps the most striking. This Western Australian native features velvety flowers and an unusual red stem.*

BELOW LEFT *The spectacular Sturt desert pea* (Clianthus formosus). *This low-trailing creeper flourishes in the dry, sandy environment of the outback and produces its brilliant red flowers, the floral emblem of South Australia, between July and January.*

BELOW RIGHT *One of the 250 varieties of grevillea which are found throughout Australia,* Grevillea rosmarinifolia *features red to creamy pink flowers arranged in orb-shaped clusters with sharp, narrow leaves. This dense shrub grows in dry environments in New South Wales and Victoria.*

WILDFLOWERS

LEFT *The almost translucent pink petals of the lotus lily (*Nelumbo nucifera*) are a common sight in Kakadu National Park during the dry season. This aquatic plant's large flowers with their yellow stamens can be as large as 30 cm across.*

RIGHT *The pink wax-flower (*Eriostemon lanceolatus*) flourishes in the coastal regions of New South Wales and southern Queensland. The shrub's modest grey-green leaves frame delicate pink five-petalled flowers which bloom between September and October each year.*

WILDFLOWERS

WILDFLOWERS

LEFT *Yellow groundsel bushes (Senecio lautus) provide a splash of colour in an otherwise muted landscape in the Willandra Lakes region of south-western New South Wales.*

RIGHT *Named after botanist, Sir Joseph Banks, the genus* Banksia *contains Australia's most identifiable flowering plants. The golden-toned Banksia robur grows to a height of up to 3 m and is just one of the 70 or so species that are unique to Australia.*

BELOW *Prevalent throughout the centre of the continent, spinifex, with its leathery, needle-shaped leaves, is a species of grass that is perfectly adapted to its arid environment.*

Reptiles, Insects and Frogs

REPTILES, INSECTS AND FROGS

TOP LEFT *The frilled lizard (Chlamydosaurus kingii) is widespread in northern Australia. This creature is famous for its impressive defence display when the large frill of skin, which normally lies close to the body, is raised to form an umbrella shape around the neck.*

LEFT *One of the many species of goanna, the sand monitor (Varanus gouldii) is Australia's most common monitor. Despite the fact that these creatures, also known as Gould's goanna, move rapidly, they are a favourite food of Aboriginal people.*

RIGHT *Australia's largest goanna, the perentie (Varanus giganteus) can reach up to 2.5 m in length. This well-camouflaged lizard inhabits the arid regions of central Australia, preying on mammals, birds and other reptiles.*

PREVIOUS PAGES *Australia's saltwater or estuarine crocodile (Crocodylus porosus) was long hunted for its skin but is now classified as an endangered species and is protected by law. This broad-snouted crocodile can grow to a length of 7 m and inhabits the coastal areas and swamps of northern Australia, feeding on fish, water birds, lizards, snakes and mammals.*

REPTILES, INSECTS AND FROGS

PICTURE CREDITS

Gunther Deichmann: page 12.

Leo Meier: front cover, back cover, half title page, opposite title page, title page,
pages 6, 11, 13 (left and right), 22, 23, 25, 27 (left), 32, 33, 34, 36, 37, 38–39, 41, 42 (left and right),
43, 46, 47, 48 (top and bottom), 53 (right), 54, 56, 57 (left), 58, 59, 60 (top and bottom), 60–61,
62, 63 (top and bottom), 64–65, 66 (top and bottom), 68–69 and 70–71.

Reg Morrison: pages 7, 14–15, 19, 24, 27 (right) and 28–29.

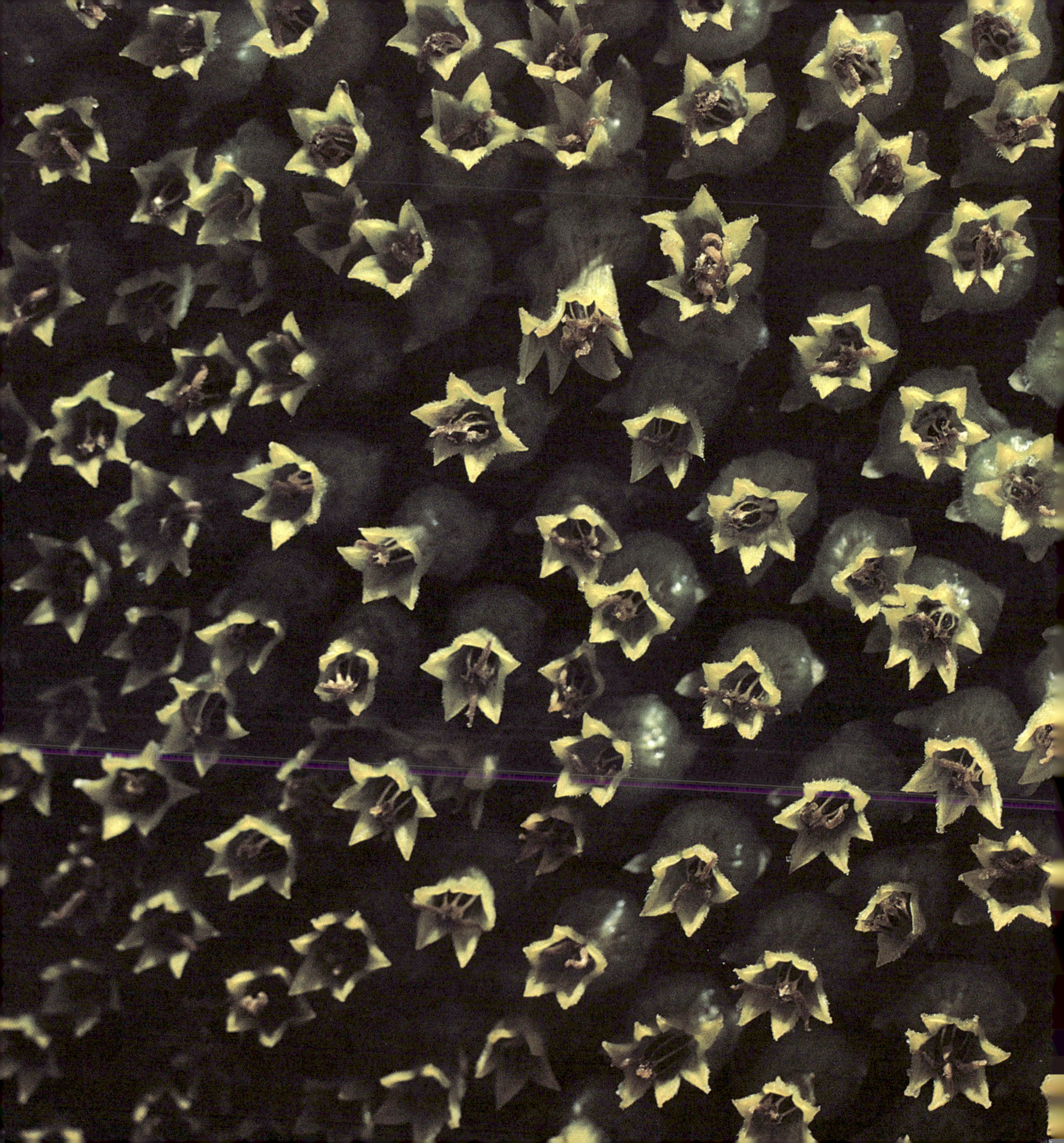